CW00327517

FERRARI
SPORTS CARS
&
PROTOTYPES

FERRARI
SPORTS CARS
&
PROTOTYPES

GIULIO SCHMIDT

Foulis

Haynes
®

© 1994 Giorgio Nada Editore, Vimodrone (Milano)

*All rights reserved. No part of this publication may be reproduced,
stored in a retrieval system, or transmitted, in any form or by any
means, electronic, mechanical, photocopying, recording
or otherwise, without prior written permission of
Giorgio Nada Editore, Vimodrone (Milan).*

First published in Italy in 1994
by Giorgio Nada Editore

This edition first published in Great Britain in 1994
by GT Foulis & Co an imprint of Haynes Publishing,
Sparkford, Nr Yeovil, Somerset, BA22 7JJ

British Library Cataloguing in Publication Data:
A catalogue record for this book is available from
the British Library

ISBN 0 85429 958 0

CONTENTS

PREFACE

Of all racing teams only Ferrari can boast twenty-three world championship titles. Nine of these belong to the single-seater Formula 1 cars, but the lion's share of this remarkable haul was won by the cars of the sports prototype category. Between 1953 (the first year of the World Constructors' Championship) and 1974, the covered-wheel *bolidi* bearing the prancing horse badge were always at the forefront and triumphed on no less than 14 occasions. These remarkable machines were responsible for memorable chapters in the history of automotive sport such as the one-two-three finish achieved by the 330s at Daytona in 1967. Enzo Ferrari was very fond of his sports cars, perhaps even more so than the Formula One single-seaters as they represented a further challenge to the titans of world car production, firms such as Ford. And, as we all know, challenges were always the most attractive aspects of life for the Drake. In the seventies, Ferrari decided to withdraw from endurance racing, the result of regulations which were no longer in tune with the firm's strategies. With the company having thus abandoned the sports prototype projects, it was up to the privateers to keep the Ferrari flag flying with production GTs suitably modified for racing.

This volume presents an anthology of the most significant sports prototypes constructed at Maranello. Our intention is to provide an overview of Ferrari sporting history through the analyses of 22 historic models from the 166 Mille Miglia to the Le Mans-prepared F40. The history of each model is complemented by a full technical specification.

166 MM

Engine: front-mounted, 60 degree V12. Displacement: 1,995.02 cc. Bore and stroke: 60 x 58.8 mm. Compression ratio: 9.5:1. Power: approx. 140 hp at 6,600 rpm. Valve gear: 2 slightly offset overhead valves per cylinder operated by one silent chain driven overhead camshaft per head. Fuel system: naturally-aspirated, diaphragm pump, 3 Weber 32DCF carburetors. Gearbox: en bloc with the engine, five speeds plus reverse. Brakes: hydraulic drums. Frame: elliptical cross section tubular steel monocoque. Suspension, front: independent, unequal length A arms, transverse lower, leaf spring. Suspension, rear: rigid live axle and longitudinal semi-elliptic leaf springs. Tracks: front, 1,270 mm; rear, 1,250 mm. Wheelbase: 2,250 mm. Weight: 680 kg.

This car belongs to the 166 generation, the first of the Ferrari sports cars, which by the end of the Forties had "diversified" into various types: S, F2, Inter, FL and MM. It was launched at the Turin Show in the November of 1948. The "MM" cypher (Mille Miglia) was chosen to commemorate Biondetti-Navone's win in the 1948 edition of that classic race. Car number 0040 M was built in 1950 and was the sixteenth of the twenty five barchettas bodied by Touring. A works car, it took part in the 1950 Giro di Sicilia-Targa Florio with Gigi Villoresi driving but failed to finish.

Then it was raced in the Mille Miglia. There was no record of who drove on that occasion (but an educated guess would be Giovanni Bracco and Umberto Maglioli who came fourth overall and first in the 2000 cc Sport clas). Subsequently, on the 20th of June 1950, 0040 was shipped out to Portugal by Jõao Gaspar, the Ferrari agent for that country. Vasco Sameiro, a Portuguese driver who raced between 1931 and 1955 in Europe and South America, drove it for the first time at

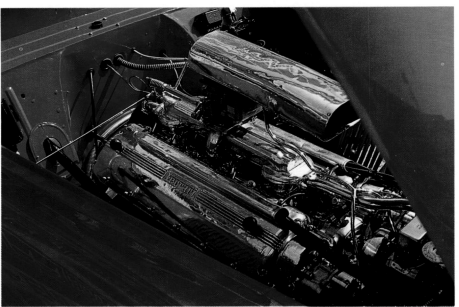

the Villa Real circuit where he took ninth place.
The car was entered for other races and was also driven by di Oliveira between 1950 and 1957. It remained in Portugal until 1973. Ownership passed from Vasco Sameiro to Manuel Palma, the Ferrari agent in Lisbon, and then to Jõao de Lacerda. Then it was bought by Derek Collins, who undertook some detailed research of the history of this particular model.

It is worth noting the curious fact that Derek Collins thought he had found a certain 166 MM with chassis number 0056 M, which had also been imported into Portugal to be raced there. During restoration work it was discovered that the car was in reality 0044 M.
The present owner is Dudley Mason-Styrron.

The most important wins for the Mk l 166 MM barchetta were the Mille Miglia with Biondetti-Salani, the Luxembourg Grand Prix Sport with Gigi Villoresi, the Le Mans 24 Hours with Chinetti-Selsdon, and the Spa 24 Hours with Chinetti-Lucas.
All these victories were achieved in 1949.

340
AMERICA LM

Engine: front-mounted 60 degree V12. Displacement: 4,101.66 cc. Bore and stroke: 80 x 68 mm. Compression ratio: 8:1. Power: 230 hp at 6,000 rpm. Valve gear: sohc, 2 overhead valves per cylinder. Fuel system: naturally-aspirated, 3 Weber 40DCF carburetors. Gearbox: five speeds plus reverse. Brakes: hydraulic drums. Frame: tubular steel side members. Suspension, front: independent, unequal length A arms, single lower transverse leaf spring. Suspension, rear: rigid live axle and longitudinal leaf springs. Tracks: front, 1,278 mm; rear, 1,250 mm. Wheelbase: 2,420 mm. Weight: 900 kg.

The 340 America LM made its first appearance at the Paris Show in October 1950. Bodied by Touring, it was a black barchetta with green leather upholstery. The styling was still that of the 166 MM, but the wheelbase was longer. Under the bonnet lurked the new 12-cylinder engine designed by Aurelio Lampredi. It was the same engine which had been fitted to the F1 car driven by Alberto Ascari at the Grand Prix des Nations in Geneva. With a displacement of 4,101.66 cc, it represented the second stage (the first was the 3.3 litre unit of the 275 S)

in the chain of development leading towards the 4.5 litres called for by the new Formula 1.

The 340 engine was to give rise to many models. As well as the 340 America (which later became the 342 America), Lampredi's new 12-cylinder engine was also fitted to the 340 Mexico and the 340 MM. Number 0122 A is one of two berlinettas which Touring prepared in the Le Mans format for its racing clientele. We know for certain that 0122 A was driven by the Blaton brothers in the 1957 Autumn Rally in Belgium. Using the 340 America as a basis, Touring built seven barchettas, Ghia four berlinettas and a 2+2 coupé, Vignale four berlinettas, five spyders, a coupé and a cabriolet. The 340 America's most important win was racked up by Gigi Villoresi driving a Vignale berlinetta in the 1951 Mille Miglia. That same year, Villoresi also won the Circuito di Senigallia event. Finally, we should note the second place taken by Vittorio Marzotto and Zanuso in the Giro della Toscana. The 340 America's Le Mans outing was less fortunate however. There was no works Ferrari lined up against the Jaguars, Talbots and Aston Martins. Of the four privately entered Touring barchettas, only the one driven by Chinetti-Lucas managed to finish. They came in eighth. Chiron-Heldé were disqualified for refuelling in advance while the clutch in Spear-Claes's car burned out. Hall-Navone, who were lying fourth at the half way stage, had to give up when the starter motor broke down.

212 MM
EXPORT SPIDER VIGNALE

Engine: front-mounted, 60 degree V12, naturally-aspirated. Displacement 2,562.51 cc. Bore and stroke: 68 x 58.8 mm. Compression ratio: 8:1. Power: 170 hp at 6,500 rpm. Valve gear: single camshaft, 2 overhead valves per cylinder. Fuel system: naturally-aspirated, 3 Weber 32DCF carburetors. Gearbox: five speeds plus reverse. Brakes: hydraulic drums. Frame: tubular steel side and cross members. Suspension, front: independent, unequal length A-arms, lower tranverse, leaf spring. Suspension, rear: double semi-elliptic leaf springs for each wheel, rigid live axle. Tracks: front, 1,270 mm; rear, 1,250 mm. Wheelbase: 2,250 mm. Weight: 1,000 kg.

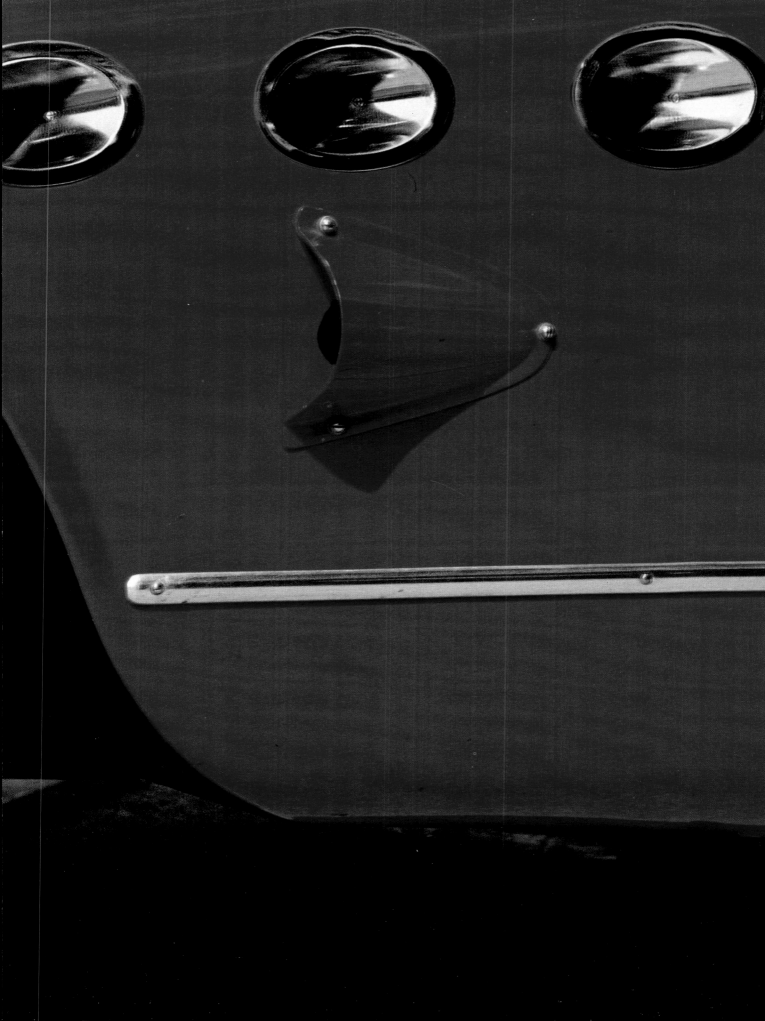

The Ferrari catalogue, distributed at the 1950 Paris Show, presented the 212 Export along with the 340 America. The chassis possessed the characteristic features of the 166 MM and the 195 S. The engine was Colombo's 12-cylinder unit which Lampredi had upgraded to 2,562.51 cc. The catalogue says nothing about the body type.

The association between Alfredo Vignale and Ferrari began with the 212 Export. Antoine Prunet recalls,

"Vignale served his time in the Farina coachworks, one of the oldest and most respected firms in Turin. In 1946 he went into business for himself and made a good job of the Cisitalia. Franco Cornacchia, the new Ferrari agent in Milan, gave him his first Ferrari, a 166 Inter."

Vignale's draughtsman was a youngster called Giovanni Michelotti. The body was hand modelled with a wooden mallet. Vignale built two cabriolets, seven berlinettas and four spyders

over 212 Export mechanics. Other coachbuilders worked with the 212 Export chassis: Motto made a spyder, Touring built seven barchettas and two berlinettas while Reggiani produced a spyder.

Ferrari's success in motorsport continued thanks to the Export and Inter versions of the 212. Jean Lucas and Jacques Peron triumphed in the 1951 Moroccan Rally driving a 212 Export berlinetta bodied by Vignale. Gigi Villoresi won the Coppa Intereuropa at Monza. Taruffi-Chinetti and Ascari-Villoresi dominated the Carrera Panamericana. The 212 Export also swept the board in the Tour de France with Pagnibon-Barraquet in a Touring bodied barchetta, Peron-Bertramnie in a Vignale berlinetta and Checcacci-Shell in a Vignale spyder. The Marzotto brothers also won numerous races: Vittorio took the Giro di Sicilia, Giannino won the Coppa della Toscana while Paolo won the Coppa d'oro delle Dolomiti.

The 212 Export MM shown on these pages is a Vignale spyder with chassis number 0214 ED.

500 MONDIAL

Engine: front-mounted, 4 cylinders in line, naturally-aspirated. Displacement: 1,984.85 cc. Bore and stroke: 90 x 78 mm. Compression ratio: 9:1 approx. Power: 155 hp at 7,000 rpm. Valve gear: dohc, 2 overhead valves per cylinder. Fuel system: naturally-aspirated, 2 Weber 42 DOE carburetors. Gearbox: rear-mounted, four speeds plus reverse. Brakes: hydraulic drums. Frame: tubular steel side and cross members. Suspension, front: independent, unequal length A-arms, transverse leaf spring. Suspension, rear: de Dion axle, transverse leaf spring. Tracks: front, 1,278 mm; rear, 1,284 mm. Wheelbase: 2,250 mm. Weight: 720 kg.

Aurelio Lampredi's masterpiece. The engine was the 2 litre, 4-cylinder unit which had powered Alberto Ascari and Ferrari to two consecutive world titles, in 1952 and 1953. This same engine turned up again inside the 500 Mondial, built to commemorate those victories. The 500 Mondial's racing debut dates from the 20th of December 1953, when Ascari-Villoresi took a laudable

second place overall and class first in the Casablanca 12 Hours. That first Mondial had been prepared by a young Modenese coachbuilder named Franco Scaglietti. As Antoine Prunet recalls, "a short while before, Dino Ferrari had designed a body which he then asked Scaglietti to adapt for an old 166 which his father had given him as a present. The result was a success, an original body

which Scaglietti was to give to numerous competition Ferraris for many years into the future." It is believed that the 500 Mondial's first outright victory dates from the 28th of March 1954 when Paolo Marzotto won the Torricelle hillclimb event. But the truth is, as Enzo Ferrari himself pointed out, that the winning car was fitted with "an experimental 4-cylinder 3 litre engine." This means that the Mondial's first real win occurred on the 9th of April 1954 when François Picard beat Jean Lucas in the 250 MM over the new Menara circuit in Marrakech. The following results are also worth noting: Vittorio Marzotto's second place overall in the 1954 Mille Miglia, the one-two scored by Umberto Maglioli and Giulio Musitelli in the Conchiglia d'oro at Imola and François Picard's win in the Penya-Rhin Grand Prix on the 23rd of October 1954. The 500 Mondial was bodied by Scaglietti and Pinin Farina. According to Godfrey Eaton, Scaglietti clothed nineteen spyders while Pinin Farina prepared three berlinettas and eleven spyders.

Two 500 Mondials were present at the Imola show organized by Ferrari Club Italia. We know that 0408 MD took part in some races in Sweden and that Franco Scaglietti rebodied it in fibreglass in 1957. Chassis number 0408 MD is not listed in Eaton's book The Complete Ferrari. 0410 MD, which boasts a yellow paint job, was present at Imola but it never left the pits. It is a spyder bodied by Pinin Farina.

750 MONZA

Engine: front-mounted 4 cylinders in line, naturally-aspirated. Displacement: 2,999.62 cc. Bore and stroke: 103 x 90 mm. Compression ratio: 9.2:1. Power: 260 hp at 6,000 rpm. Valve gear: dohc, 2 overhead valves per cylinder. Fuel system: naturally-aspirated, 2 Weber 58 DCO A/3 carburetors. Gearbox: rear-mounted, five speeds plus reverse. Brakes: hydraulic drums. Frame: tubular steel side and cross members. Suspension, front: independent, unequal length A-arms, coil springs. Suspension, rear: de Dion axle, transverse leaf spring. Tracks: front, 1,278 mm; rear, 1,294 mm. Wheelbase: 2,250 mm. Weight: 720 kg.

According to some historians, the 750 Monza did not in fact make its debut, as was hitherto believed, at the Supercortemaggiore Grand Prix held at Monza on the 27th of June 1954, but at the Reims 12 Hours of July 3rd 1954, with Maglioli-Manzon. In the 1954 Annuario (Yearbook), Enzo Ferrari stated: "The Supercortemaggiore Grand Prix witnessed the official debut of the 3 litre, 4-cylinder

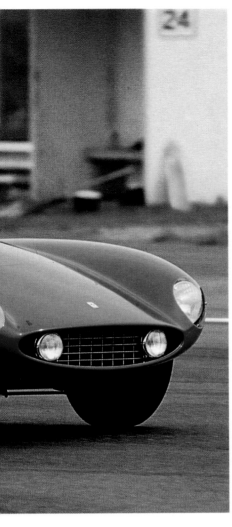

car which was raced by works Ferrari drivers..." More specifically Mike Hawthorn-Umberto Maglioli and José Froilan Gonzalez-Maurice Trintignant. Ferrari added: "There were many different types of cars in the race; including various Ferraris. For example, the 3 litre cars driven by Gerini-Cornacchia and Biondetti-Nocentini belonged to the type which immediately preceded the Monza." He went on to explain that Paolo Marzotto's win in the Torricelle hillclimb, an event held on the 28th of March 1954 near Verona, was achieved with an experimental car fitted with a 3 litre, 4-cylinder engine, i.e. a forerunner of the 750 Monza. One thing is certain however, the 750 Monza was an instant success. Hawthorn-Maglioli's win in the Supercortemaggiore Grand Prix at Monza with Gonzalez-Trintignant coming second was a good omen. Ferrari won its second World Championship in 1954 thanks to Hawthorn-Trintignant in the 750 Monza.

They dominated the Tourist Trophy and were placed second due to the regulations governing handicap races. Car number 0520 M, which is shown on these pages, was bought by Louis Rosier in 1955. The French driver took this car, a Scaglietti spyder, to the Dakar Grand Prix of the 13th March 1955 where he took second place overall. Rosier won the Forez event at St. Etienne on the 8th of May and the Bougie Grand Prix in Algeria, held on the 30th of May. Finally, he took sixth place overall in the Portuguese Grand Prix of the 26th of June. Today the car is owned by Jean Sage.

According to Godfrey Eaton at least thirty-three 750 Monzas were built. All of them were bodied by Scaglietti, with the exception of a spyder (0486 M) bodied by Pininfarina.

250 MONZA

Engine: front-mounted 60 degree V12, naturally-aspirated. Displacement: 2,963.45 cc Bore and stroke: 68 x 68 mm. Compression ratio: 9:1. Power: 240 hp at 7,200 rpm. Valve gear: sohc, 2 overhead valves per cylinder. Fuel system: naturally-aspirated, 3 Weber 36DCF/3 carburetors. Gearbox: rear-mounted, four speeds plus reverse. Brakes: hydraulic drums. Frame: tubular steel side and cross members. Suspension, front: independent, unequal length A-arms, leaf spring. Suspension, rear: de Dion axle, transverse leaf spring. Tracks: front, 1,300 mm; rear, 1,320 mm. Wheelbase: 2,400 mm. Weight: 850 kg.

Enzo Ferrari once recalled: "The new 12-cylinder, 3 litre engine had its first outing at the Hyères 12 Hours where, true to an established tradition, it enjoyed a victorious debut. Trintignant and Piotti, as well as taking first place overall, also headed the final rankings based on the index of performance, an unusual and difficult feat when driving a car with a large engine."

Trintignant-Piotti's 250 Monza had been prepared by Pinin Farina, who had lengthened the chassis of the 750 Monza before fitting it with a 250 MM engine. The body was similar to that of the 375 MM spyder and identical to that of the 500 Mondial, with the exception of the bonnet length. Four of these were built, all of them spyders: two were by Pinin Farina (0420 M and 0432 M), and two by

Scaglietti (0442 M and 0466 M). Two cars were present at the Imola Show: 0432 M and 0466 M. It is interesting to note that 0432 M, certainly the work of Pinin Farina, was rebodied by Scaglietti at the end of the Fifties. Apart from the eleventh place overall taken by Piotti-Manzon at the Supercortemaggiore Grand Prix at Monza and Piotti's win at the Circuito di Reggio Calabria, both of which results were obtained in 1954, 0432 M also took part in the Mille Miglia of the following year with "Kammamuri" (the pseudonym of Erasmo

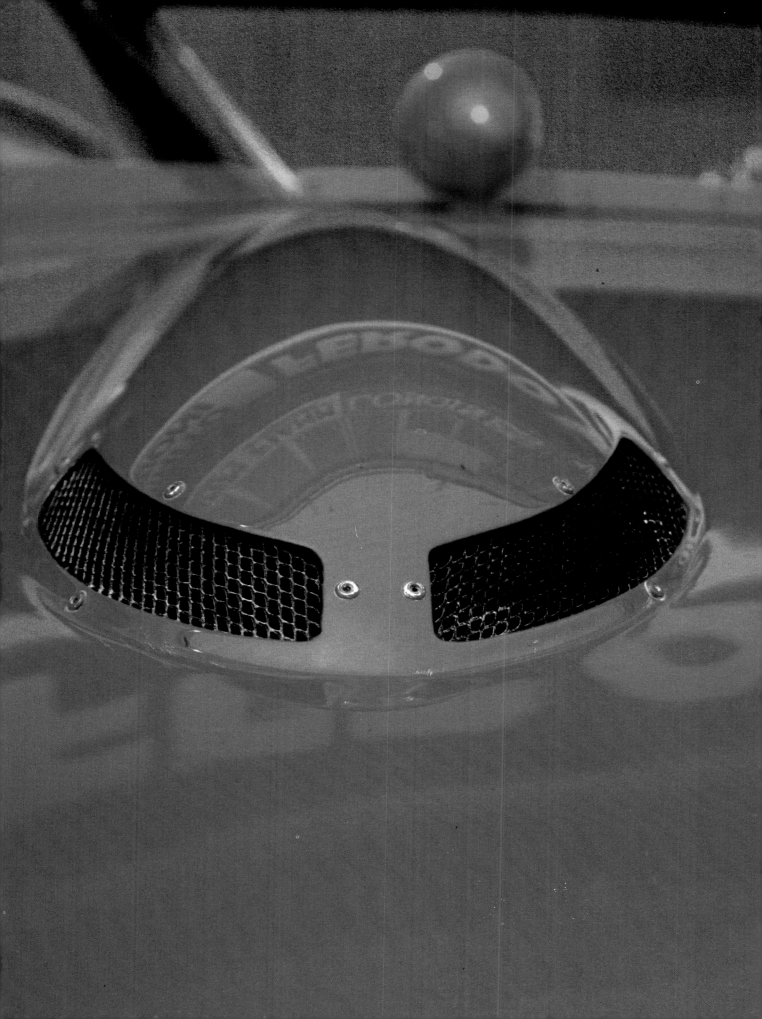

Simeoni), who came in fifteenth overall. The present owner, Peter G. Sachs, told us that the car raced in Argentina from 1956 to 1958 after which it "emigrated" to the USA where it remained for 18 years.

Franco Lombardi's 0466 M was the last of the four 250 Monzas produced in 1954. A works car, it won the 1954 Senigallia Grand Prix with Umberto Maglioli. Compared to the other 250 Monzas, 0466 M was a notably improved version as Ferrari had used it as a mobile test bed. It was sold in 1955 to the Guastalla racing team run by Franco Cornacchia and Gerino Gerini, and then to the Fayen-Dos Santos partnership who raced the car in the 1956 Buenos Aires 1000 Kilometres, where it failed to finish after starting in seventh place. Finally, the car spent some time in Venezuela before returning to Italy in 1978.

860 MONZA

Engine: front-mounted 4 cylinders in line, naturally-aspirated. Displacement: 3,431.93 cc. Bore and stroke: 102 x 105 mm. Compression ratio: 8.5:1. Power: 280 hp at 5,800 rpm. Valve gear: dohc, 2 overhead valves per cylinder. Fuel system: naturally-aspirated, 2 Weber 58DCO/A carburetors. Gearbox: rear-mounted, four speeds plus reverse. Brakes: hydraulic drums. Frame: tubular steel side and cross members. Suspension, front: independent, unequal length A-arms, coil springs. Suspension, rear: 4 radius arms with transverse leaf spring, hydraulic shock absorbers. Tracks: front, 1,378 mm; rear, 1,324 mm. Wheelbase: 2,300 mm. Weight: 860 kg.

The unlucky 1955 season, in which Mercedes won virtually everything, was a thing of the past. Public opinion was still shaken by the death of Alberto Ascari and Lancia's withdrawal from racing. Enzo Ferrari rolled up his sleeves and prepared to make his comeback.

Lined up by Ferrari alongside the 12-cylinder 290 MM, the 4-cylinder, 3.5 litre 860 Monza made an important contribution to the conquest of Maranello's third world title in 1956. It scored a thrilling one-two at the Sebring 12 Hours with Fangio-Castellotti and Musso-Schell. It came in behind Castellotti's 290 MM in the Mille Miglia, with Collins and Musso second and third respectively. It won again at the Nürburgring in the 1000 Kilometres with Fangio-Castellotti, while Hawthorn-De Portago-Hamilton took a precious third place in the Swedish Grand Prix, the last event on the World Championship card. Outwith the championship, it won the Rouen Grand Prix with Eugenio Castellotti driving.

Towards the end of 1955, the car made its debut in the Tourist Trophy with the denomination 857 S. The story of the 860 Monza which appears on these pages, chassis number 0578 MD, is told in a letter which Jess G. Pourret sent to its owner in 1984. "It is exactly the same vehicle which I rediscovered in the Bronx: a 1955 car which had been a 4-cylinder, 3 litre 750 Monza subsequently converted by the manufacturer into an 860 Monza (857 S type). It was certainly the first 860 Monza of 1955. Then it was sold to Alfonso De Portago. Its racing career came to an abrupt end in 1958 after it emerged somewhat the worse for wear after a serious crash." We can add that the 750 Monza referred to by Pourret had been in its turn derived from a 500 Mondial (the MD cypher following the chassis number shows this). It took part in numerous races with Alfonso De Portago and Phil Hill.

According to Godfrey Eaton, no more than seven 860 Monzas were built.

290 MM

Engine: front-mounted 60 degree V12, naturally-aspirated. Displacement: 3,490.61 cc. Bore and stroke: 73 x 69.5 mm. Compression ratio: 9:1. Power: 320 hp at 7,300 rpm. Valve gear: sohc, 2 overhead valves per cylinder. Fuel system: naturally-aspirated, 3 Weber 36 IR4/C1 carburetors. Gearbox: rear-mounted, four speeds plus reverse. Brakes: hydraulic drums. Frame: tubular with punt. Suspension, front: independent, unequal length A-arms, coil springs. Suspension, rear: de Dion axle, transverse leaf spring. Tracks: front, 1,310 mm; rear, 1,286 mm. Wheelbase: 2,350 mm. Weight: 880 kg.

The 290 MM marked the return to the 12-cylinder format. This was a new engine of a markedly different design from the Colombo and Lampredi versions. This was the time of Vittorio Bellentani, and Alberto Massimino, who had created the Alfa Romeo 158 along with Gioachino Colombo, and a very young engine specialist called Andrea Fraschetti. This trio worked under the influence of Vittorio Jano, who had come back to work for Enzo Ferrari as a consultant following Gianni Lancia's decision to withdraw from motorsport. The 290 MM's first race was the Giro di Sicilia on the 8th of April 1956. It was an unlucky debut because Castellotti and Musso, whose cars were fitted with short exhausts and large air intakes on the bonnets, both failed to finish. The 290 MM made a decisive contribution to Ferrari's World Championship victory in 1956. It won two

championship events, the Mille Miglia with Castellotti and the Swedish Grand Prix with Trintignant-Phil Hill. There was another 290 MM at the Mille Miglia, driven by Juan Manuel Fangio, who had to be content with fourth place. At the Nürburgring

1000 Kilometres, Phil Hill and Ken Wharton, aided by Gendebien and De Portago, took third place in the 290 MM, while the car driven by Musso-Trintignant withdrew after a crash. At the Swedish Grand Prix, another 12-cylinder Ferrari with von Trips-Collins came in second behind winners Trintignant and Phil Hill in the 290 MM. The 290 MM on show at Imola was number 0628. This was the car with which Castellotti triumphed in the 1956 Mille Miglia. The following year, the same car won the Buenos Aires 1000 Kilometres with Masten Gregory, Cesare Perdisa, Eugenio Castellotti and Luigi Musso. It failed to finish at the Sebring 12 Hours with Phil Hill and von Trips at the wheel.
At the end of 1957 it bowed out in grand style after winning in Cuba and Nassau with drivers Masten Gregory and Stirling Moss respectively.

500 TRC

Engine: 4 cylinders in line, front-mounted, naturally-aspirated. Displacement: 1,984.85 cc. Bore and stroke: 90 x 78 mm. Compression ratio: 9.1:1. Power: 190 hp at 7,400 rpm. Valve gear: dohc, 2 overhead valves per cylinder. Fuel system: naturally-aspirated, 2 Weber 40DCO/A3 carburetors. Gearbox: front-mounted, four speeds plus reverse. Brakes: hydraulic drums. Frame: tubular steel side and cross members. Suspension, front: independent, unequal length A-arms, coil springs. Suspension, rear: rigid live axle, coil springs. Tracks: front, 1,308 mm; rear, 1,250 mm. Wheelbase 2,250 mm. Weight: 680 kg.

This represents the evolution of the 2 litre, 4-cylinder Sports series which began with the 500 Mondial before leading to the 500 TR and finally the 500 TRC, intended for the racing clientele. It was built in accordance with the formula laid down in appendix "C" of the FA regulations for the 1957 season. The body, designed by Pinin Farina, was built by Scaglietti. It was 100 mm lower than the 500 TR, with an extremely ample

windscreen and a passenger door on the left hand side. The car also had a soft top. Gino Munaron observed: "This latter accessory, still required by appendix C, was decidedly funny looking, ugly even, and of no use whatsoever: the air pressure even at 60/70 km/h would have torn it to ribbons. The soft top had to be put up for examination during the regulation technical checks. Afterwards it was not only

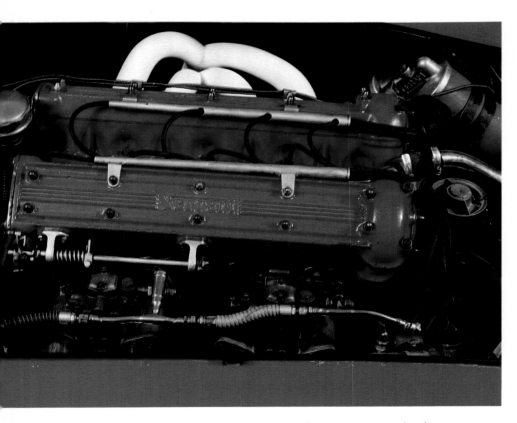

2 litre class. It was the last 4-cylinder Sports car built by Ferrari.

Two 500 TRCs took part in the Imola show: 0668 MDTR and 0702 MDTR.

Luigi Chinetti sold car number 0668 in the United States, where it was entered in numerous races. Number 0702 raced for a while in the colours of the Belgian Francorchamps team. It seems that it took part in the 1957 Le Mans 24 Hours, since the works car earmarked for Lucien Bianchi and Harris had been destroyed in a smash before the race. It came in seventh. In the Swedish Grand Prix, the 500 TRCs dominated the up to 2000 cc class taking the first three places: 0702 was probably driven by Harris-Liekens, who came in third.

In 1958 the car was sold to Chinetti's NART stable where the car was driven by William Cooper.

possible to put it back down again - you could take the thing off altogether and leave it at home or in the pits." It was never raced as a works car, but it was a winner just the same. It picked up handsome class wins at circuits all over the world. Production ceased in 1958 when the new 6-cylinder Dino took over from the TRC in the

250 TR

This brings us up to the time when the "monsters" were banned. The tragedy at Le Mans was still an open wound. For the 1958 season, the World Championship was reserved for cars of up to 3 litres capacity. Ferrari was not caught unprepared. On the 26th of May 1957, at the

Nürburgring 1000 Kilometres, a prototype 250 TR fitted with a 12-cylinder twin cam engine made its appearance. Despite a modest placing -Masten Gregory and Olindo Morolli had to be content with tenth place -the experience was considered a positive one. Another prototype, with an engine directly derived from the 250 GT bolted onto a 500 TRC chassis, did very well at the Le Mans 24 Hours, before it was obliged to withdraw. After a disappointing show in the Swedish Grand Prix, the future 250 TRs took third and fourth places in the last event valid for the World

Engine: front-mounted 60 degree V12, naturally-aspirated. Displacement: 2,953.21 cc. Bore and stroke: 73 x 58.8 mm. Compression ratio: 9.8:1. Power: 290 hp at 7,500 rpm. Valve gear: sohc, 2 overhead valves per cylinder. Fuel system: naturally-aspirated 6 Weber 38 DCN carburetors. Gearbox: front-mounted, four speeds plus reverse. Brakes: hydraulic drums. Frame: tubular space-frame. Suspension, front: independent, unequal length A-arms, coil springs. Suspension, rear: de Dion axle, coil springs. Tracks: front, 1,308 mm; rear, 1,300 mm. Wheelbase: 2,350 mm. Weight: 800 kg.

Championship, the Venezuelan Grand Prix in Caracas, where von Trips-Seidel and Trintignant-Gendebien finished right behind the 4 litre Ferraris driven by Collins-Phil Hill and Hawthorn-Musso. On the 22nd of November, finally, Ferrari announced the launching of the new 250 TR, which he was preparing to spearhead his 1958 title challenge. Two versions: a standard model for the clients and one for the works drivers. The engine, derived from the 166 and first used to power the GTs and then the Sports cars, was to be the same for both versions. Its preparation had been supervised by Carlo Chiti, who had joined Ferrari following the death of Andrea Fraschetti, killed while test driving a car. The differences between the two versions lay in the frame: there was a rigid back axle for the standard version while a sophisticated de Dion axle had been fitted to the works cars. The 250 TR dominated the 1958 season, threatened only by Aston Martin and the "little" Porsche. It won the Buenos Aires 1000 Kilometres and the Sebring 12 Hours with Collins-Phil Hill, the Targa Florio with Musso-Gendebien and the Le Mans 24 Hours with Gendebien-Phil Hill. In the Nürburgring 1000 Kilometres, won by Aston Martin, it took second, third, fourth and fifth place. In the final rankings of the World Manufacturers' Championship, Ferrari had a total of 38 points, Porsche 19 and Aston Martin 18. The 250 TR underwent further development: in 1959, it was a disappointment; in 1960 and 1961, it presented Ferrari with two new world titles. The 250 TR ended its career as a works car in 1961. Number 0758, the car shown on these pages, was sold in the standard version to the Lagartixa racing team in 1958. It raced exclusively in Venezuela. José Miguel Galia won the Vuelta di Aragna three times in a row between 1958 and 1960 in this car. He also won the Circuito Cumbres de Curumo event in 1960. A photograph of Galia in action over this latter circuit was published in the 1960 Annuario Ferrari.

45

250 LE MANS

Engine: rear-mounted 60 degree V12, naturally-aspirated. Displacement: 3,285.72 cc. Bore and stroke: 77 x 58.8 mm. Compression ratio: 9.7:1. Power: 320 hp at 7,500 rpm. Valve gear: sohc, 2 overhead valves per cylinder. Fuel system: naturally-aspirated, 6 Weber 38 DCN or 40 DCN 2 carburetors. Gearbox: rear-mounted, five speeds plus reverse. Brakes: discs. Frame: tubular space-frame. Suspension, front: independent, wishbones, coil springs, anti-roll bar. Suspension, rear: independent, wishbones, coil springs, anti-roll bar. Tracks: front, 1,350 mm; rear, 1,340 mm. Wheelbase: 2,400 mm. Weight: 20 kg.

result was a GT berlinetta version of the spider Sport. In the 1964 Annuario Ferrari (Ferrari Yearbook) there is a significant sentence: "The 250 Le Mans, a car for people who like to drive fast and who know how to do it." Mauro Forghieri gave us a valuable nugget of information when he told us that the 250 LM was the first

Ferrari to be tested in a wind tunnel, at the MIRA establishment in England. Mike Parkes tested it on the road and the track. It was the fruit of a team effort on the part of Ferrari Racing and the Production division. Designed by Pininfarina and bodied by Scaglietti, the first Le Mans to be produced had a 3 litre, 12-cylinder engine: that was why it was called the 250, after the displacement of the individual cylinders. All the other LMs produced were fitted with the 3.3 litre 275 engine. Despite this, they left the name unchanged.

David Piper observed: "The 250 Le Mans has an important place in Ferrari history because it was the first rear-engined berlinetta ever built at Maranello and, with the exception of the sports cars prototypes, it was also the first car of this type to be sold to private individuals. Although the Le Mans was not a highly innovational car technologically speaking, it is nonetheless true that it represented the beginnings of a philosophy which may still be discerned in the methods of construction currently in use at Ferrari.

"After the success of the 250 P, Ferrari decided to make the 250 Le Mans berlinetta for his racing clientele. The

The World Manufacturers' Championship, which until 1963 had been fought out by sports cars, was reserved for GTs in 1964. Unfortunately a considerable time was to pass before the Le Mans received its GT homologation and so the car was obliged to race in the Prototype class. Despite this, it made an honourable showing all over the world. David Piper purchased car number 8165 in the June of 1966, after the Filipinetti team had entered it for the Nürburgring 1000 Kilometre.

It has a fiberglass body in preference to aluminum and Piper had numerous modifications made in order

to make the car as competitive as possible. The car has taken part in fourteen races altogether. David Piper has won outright on four occasions: the Guards Trophy at Brands Hatch, the Oulton Park Gold Cup and the Paris 1000 Kilometres (in which Piper was teamed with Mike Parkes) in 1966, and the Daily Express Trophy at Silverstone in 1967.

There was also a class win in the 1967 Brands Hatch 500 Miles event with Dibley-Pierpoint. On other occasions the car was driven by Dick Attwood, Brian Redman, Paul Vestey, Lucien Bianchi and Pedro Rodriguez.

330 P

Engine: rear-mounted 60 degree V12, naturally-aspirated. Displacement: 3,285.72 cc. Bore and stroke: 77 x 58.8 mm. Compression ratio: 9.7:1. Power: 320 hp at 7,500 rpm. Valve gear: sohc, 2 overhead valves per cylinder. Fuel system: naturally-aspirated, 6 Weber 38 DCN or 40 DCN 2 carburetors. Gearbox: rear-mounted, five speeds plus reverse. Brakes: discs. Frame: tubular space-frame. Suspension, front: independent, wishbones, coil springs, anti-roll bar. Suspension, rear: independent, wishbones, coil springs, anti-roll bar. Tracks: front, 1,350 mm; rear, 1,340 mm. Wheelbase: 2,400 mm. Weight: 20 kg.

The 275 P's big sister, both cars were descended from the 250 P. Of an age with the 250 LM, the 330 P has a 4 litre, 12 cylinder engine while the 275 P has a 3.3 litre unit. The power varied: the 330 P generated 370 hp, the 275 P produced 320 hp. The 330 engine is the Testa Rossa version of the 400 SA. Back in 1963, at the Reims Trophy of the 30th of June, Ferrari had already played the 4 litre P card. On that occasion, Mike Parkes beat the lap record more than once and only clutch failure robbed the English engineer-driver of victory. Another experiment was carried out at Silverstone in the sports car race which precedes the Formula 1 Grand Prix. Mike Parkes was unlucky again. The 330 P was involved in a "shunt" and was left

unable to display her talents. In 1964 three new chassis were made ready (0818, 0820 and 0822). The bonnet (which opened up towards the nose) of the new body resembled that of the 250 LM. These three frames housed either the 275

engine or the 330 unit. Powered by the 275 engine, number 0820 took second place in the Sebring 12 Hours with Scarfiotti-Vaccarella and it won the Nürburgring 1000 Kilometres, again with Scarfiotti-Vaccarella. Chassis

number 0820 corresponds to the 275 P driven by Parkes and Scarfiotti which took part in the 1964 Le Mans 24 Hours but failed to finish thanks to a broken piston.

With the more powerful 4 litre engine, the 330 P was much faster, but this engine was not used much by the works drivers who preferred the car with the 275 unit.

Nevertheless, when driven by certain illustrious clients, it scored some important wins. Graham Hill won the Tourist Trophy, Pedro Rodriguez took the Canadian Grand Prix, Graham Hill and Joachim Bonnier were first in the Paris 1000 Kilometres and Lodovico Scarfiotti triumphed in the Bettoja Trophy. In the 1965 Sebring 12 Hours, number 0820, which had been acquired by the Texan Mecom Racing team, was driven by Graham Hill in tandem with Rodriguez. They were up with the leaders until the ninth hour when technical problems obliged them to withdraw. The 1964 season was full of rich satisfactions for Ferrari as they won the Formula 1 World Championship with John Surtees as well as the World Manufacturers' Championship.

365 P (P2)

Engine: rear-mounted 60 degree V12, naturally-aspirated. Displacement: 4,390.35 cc. Bore and stroke: 77 x 71 mm. Compression ratio: 9:1. Power: 380 hp at 7,300 rpm. Valve gear: sohc, 2 overhead valves per cylinder. Fuel system: naturally-aspirated, 6 Weber 42DCN carburetors. Gearbox: rear, five speeds plus reverse. Brakes: discs. Frame: tubular space-frame braced with sheet metal. Suspension: independent, unequal length A-arms, coil springs. Tracks: front, 1,400 mm; rear, 1,370 mm. Wheelbase: 2,400 mm. Weight: 860 kg.

The 365 P is a P2 with a 4.4 litre engine, obtained by reboring the 330 P single cam engine used in 1964. It was supplied to the various privately run racing teams which supported Ferrari in international events. The P2 was used by the Filipinetti team, NART, the Francorchamps team and Maranello Concessionaires. The first 365 P bore chassis number 0824. It was delivered to the Filipinetti team in the April of 1965, just before practice was scheduled to begin for the Le Mans 24 Hours. This car is worthy of note because it was the only one which was to keep the 1964 chassis and body. Ferrari had prepared six new frames for the 1965 season. Five of these, including the car shown on

David Piper
AUTO RACING

DAVID

16

MAGNETI
MARELLI
pininfarina

MAR
COLLECT

these pages (number 0838), were equipped by the makers with either 3.3 or 4 litre engines, depending upon circumstances, before being definitively converted into 365 Ps. The sixth and last model (0838) was a 365 P right from the start. Number 0836, which you can see here, enjoyed a brilliant racing career. In 1965, Parkes-Guichet won the Monza 1000 Kilometres, using the 3.3 litre engine. Towards the end of the season the car was sold to David Piper. The English driver had her converted into a sports car which he raced on the 16th of September at Mount Tremblant in Canada, where he was runner up behind John Surtees in a Lola. A fortnight later, he came fifth in the Canadian Grand Prix at Mosport. Then Piper landed in South Africa. There he won the Kyalami 9 Hours and the Angola Grand Prix.

In 1966 he had numerous modifications made to his private P2, using technical solutions "pirated" from the works version. In this way the car remained highly competitive. It won the Auvergne Trophy at Clermont Ferrand, the Kyalami 9 Hours and the Cape 3 Hours; it came fifth at Snetterton in the Scott Brown Trophy and ninth in the Tourist Trophy at Oulton Park. In 1967, it did not finish at Daytona and Sebring, but it ran out the winner in the Reims 12 Hours and took third place in the Wills Trophy at Croft.

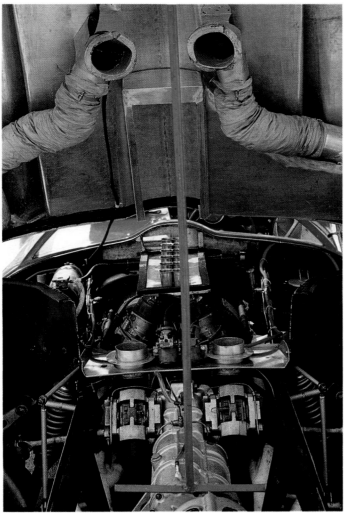

DINO 206 S

Engine: rear-mounted 65 degree V6, naturally-aspirated. Displacement: 1,986.61 cc. Bore and stroke: 86 x 57 mm. Compression ratio: 11:1. Power: 220 hp at 9,000 rpm. Valve gear: sohc, 2/3 overhead valves per cylinder. Fuel system: Lucas injection. Gearbox: rear-mounted, five speeds plus reverse. Brakes: discs. Frame: tubular space-frame braced with sheet metal. Suspension: independent, unequal length A-arms, coil springs. Tracks: front, 1,360 mm; rear, 1,355 mm. Wheelbase: 2,280 mm. Weight: 580 kg.

This is a member of the 6 cylinder Dino family. Its arrival was announced at the same time as that of the 330 P3. In 1966, it took part in several track events. Designed as a sports car, it was obliged to race in the Prototype category. Union agitation had resulted in no more than sixteen of the minimum requirement of fifty cars being built. The body, built by Pietro Drogo, resembles that of the 330 P3.

The car made its debut on the 26th of March 1966 at the Sebring 12 Hours. Driven by Bandini-Scarfiotti, it was the only 206 S present and took fifth place overall. Four Dinos were entered for the Monza 1000 Kilometres: a berlinetta injection driven by Bandini-Scarfiotti; a Maranello Concessionnaires spyder for Piper-Attwood; a spyder too for Biscaldi-Casoni of the Sant Ambroeus team; and finally

another berlinetta for Bondurant-Vaccarella. This last was damaged during practice and did not race. Of the three Dinos which did race, the best result was Bandini-Scarfiotti's tenth place. All three cars had problems with the windscreen wipers. Things went better at the Targa Florio, where Guichet-Baghetti's berlinetta, with carburetors and dual ignition, took first place. The 206 S gave of its best at the

Nürburgring 1000 Kilometres. Runners-up Scarfiotti and Bandini in the works spyder with twin-spark ignition and fuel injection came in behind Phil Hill and Joachim Bonnier in the Chaparral: a prestigious result. The excellence of the Dino 206 S was underlined by the third place taken by Pedro Rodriguez and Richie Ginther driving the carbureted spyder. On the other hand, the Le Mans showing was disastrous. None of the three 206 Ss in the race managed to finish. The first wins were racked up by private entrants. Mike Parkes won the 2 litres class at Brands Hatch. "Pam", better known as Marcello Pasotti, won the Coppa Città di Enna. The 206 S won several hillclimb events. However, Lodovico Scarfiotti was unable to hold on to the title he had won the year before.

There were two Dino 206 S at the Imola show. Number 026 is a works car which was sold to the Filipinetti team in 1967. It took part, without success, in the Sebring 12 Hours with Müller-Klass and was badly damaged during practice for the Nürburgring 1000 Kilometres. Number 030 was driven by amateur drivers in numerous races.

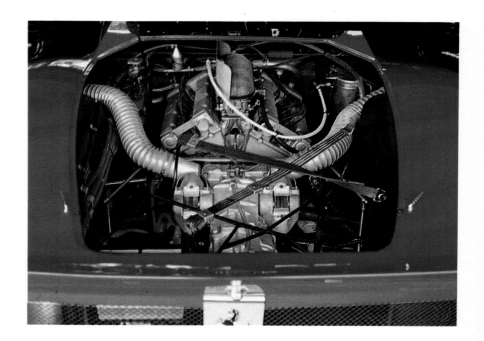

330 P4

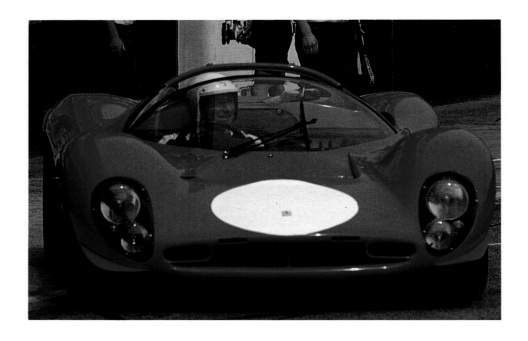

Engine: rear-mounted 60 degree V12, naturally-aspirated. Displacement: 3,967.44 cc. Bore and stroke: 77 x 71 mm. Compression ratio: 11:1. Power: 450 hp at 8,200 rpm. Valve gear: dohc, 3 overhead valves per cylinder. Fuel system: Lucas injection. Gearbox: five speeds plus reverse. Brakes: discs. Frame: tubular space-frame braced with sheet metal. Suspension: independent, unequal length A-arms, coil springs. Tracks: front, 1,488 mm; rear, 1,450 mm. Wheelbase: 2,400 mm. Weight: 664 kg.

The 330 P4 first took to the track at the beginning of the 1967 season and it soon took revenge for the 330 P3's failure to cope with the Fords the year before. The engine: that same 12-cylinder, 36 valve unit which had given Scarfiotti his only Formula 1 victory at Monza in the 1965 Italian Grand Prix. Power: the P3's 330 hp had been upgraded to 450 in the P4. There was also a new Ferrari-built gearbox, and new Campagnolo rims. The latter were made to a star shaped design which became a characteristic Ferrari feature from then on. In shape the P4 was very like the P3, but there were some differences. The P4's bonnet and radiator grille are a little flatter; the streamlining of the headlights is a little less noticeable and the vents in the vertical panel of the stubby tail are a different shape. The P4 was also longer, taller and wider than its predecessor. The car was already ready by the November of 1966. Initial testing got underway at

Daytona that December, in preparation for the 24 Hours and the showdown with Ford. The testing was immensely useful. On the 4th and 5th of February 1967, three Ferraris streaked over the finishing line to sweep the board in a triumphal victory (this arrival was dubbed a "Lini," because the new racing manager, the journalist Franco Lini, had

ordered it). Two P4s took the first two places with Amon-Bandini and Parkes-Scarfiotti while Rodriguez-Guichet were third in a 412 P. Ferrari had humbled Ford on the latter's home ground. After that, the two rival manufacturers sought to avoid each other. Ferrari abandoned Sebring. Ford turned its back on Monza where Bandini-Amon and Scarfiotti-Parkes drove P4s to the first two places in the 1000 Kilometres event. Then Fortune suddenly frowned on Ferrari. Lorenzo Bandini's death was followed by a loss of form. Parkes and Scarfiotti, in the only P4 present at Spa, came in fifth while victory fell to the Mirage-Ford which was more or less on its first outing. A curb stone foiled the P4 driven by Vaccarella-Scarfiotti in the Targa Florio. Things got better at Le Mans where P4s driven by Scarfiotti-Parkes and Mairesse-Beurlys came second and third behind Ford. The other P4s in the race, those with Klass-Sutcliffe and Amon-Vaccarella, both failed to

finish. Despite the defeat at Le Mans, Ferrari won the World Prototype Championship, thanks largely to the fact that Amon-Stewart, Scarfiotti-Sutcliffe and Williams-Hawkins took second, fifth and seventh place

respectively in the Brands Hatch 500 Kilometres. Not only had Ferrari beaten Ford, it had also beaten Porsche, which had emerged as a far more dangerous rival than the American manufacturer. Car number 0856, shown on

these pages, was driven by Parkes and Scarfiotti in the Daytona 24 Hours and the Spa 1000 Kilometres, by Mairesse-Beurlys in the Le Mans 24 Hours and by Scarfiotti-Sutcliffe in the Brands Hatch 500 Hours.

512 M

Engine: 60 degree V12, naturally-aspirated. Displacement: 4993.53 cc. Bore and stroke: 87 x 70 mm. Compression ratio: 11.8:1. Power: 600 hp at 8,000 rpm. Valve gear: dohc, 4 overhead valves per cylinder. Fuel system: Lucas injection. Gearbox: five speeds plus reverse. Brakes: discs. Frame: semi-monocoque. Suspension: independent, unequal length A-arms, coil springs. Tracks: front, 1,518 mm; rear, 1,511 mm. Wheelbase: 2,400 mm. Weight: 840 kg.

A 512 S derivative, this car was defined by Ferrari as "an absurdity required by anachronistic racing regulations." For the 1970 season, the CSI had imposed the construction of twenty-five cars as the homologation requirement for the 5 litre Sportscar class. Ferrari, having received concrete backing from Fiat, had no difficulty in building the required number in order to mount its challenge. The 512 M only won once race valid for the World Championship: the Sebring 12 Hours with Giunti-Vaccarella-Andretti. At the end of the 1970 season a refurbished 512, called the "M," won the Kyalami 9 Hours.

Car number 1040 is an original interpretation of the 512 M. Its story begins with one Kirk F. White, a Pennsylvania entrepreneur who decided to enter the 1971 Manufacturers' Championship in order to promote his business. Having no racing experience, he availed himself of the expert

services of Roger Penske, who agreed to prepare a Ferrari 512 in his workshop in Newton Square. The car in question had been acquired in August 1970 by two Californians, Chris Cord and Steve Earle. Mark Donohue and David Hobbs were called in to drive this new "American Ferrari." Roger Penske invited one of America's most

talented specialist mechanics, Lujie Lesovski, to join the team. Then he got Berry Plasti-Glass of Los Angeles to make a new body. Two engines, bought from Ferrari, were put into the hands of Traco Engineering of Culver City, California, who squeezed 600 hp out of one and 630 out of the other. After the first tests were held on the track at

Daytona, Donohue and Don Cox, an engineer with Penske Racing, involved Sunoco in the construction of an Indianapolis type refuelling system. The advantages of this were apparent right from the first race, the 1971 Daytona 24 Hours. When Donohue went into the pits for the first refuelling stop he held a lead of 1.5 seconds over Pedro

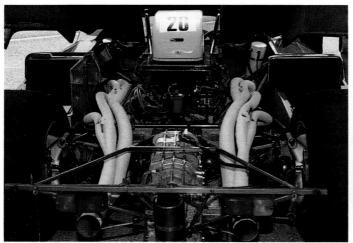

Rodriguez and Jack Oliver in the Porsche 917; when he left the pits the gap had increased to 8 seconds. The "American Ferrari," with its blue and yellow paintwork, was a contender in the 1971 season, but it never managed to win a race.

Technical problems, often rather silly ones, resulted in the car failing to finish on many occasions. Its sporting record is as follows. The Daytona 24 Hours: Donohue-Hobbs, best time in practice, third. The Sebring 12 Hours: again best in practice, Donohue-Hobbs came sixth in the race after Pedro Rodriguez in the Porsche collided with them. The Le Mans 24 Hours: after four hours it was lying second behind Rodriguez-Oliver in the Porsche; a few minutes later, the engine blew up and that was that. The Watkins Glen 6 Hours: yet again the fastest in practice and the unluckiest during the race; the suspension broke while the car was in the lead.

308 GTB/4 CARMA

Ferrari withdrew from endurance racing for some years in order to concentrate on the Formula 1 championship. But "private" drivers and specialist car preparers carried on racing in the World Championship. In 1981 the 308 GTB/4 Carma prepared by Carlo Facetti and Martino Finotto called particular attention to itself. Nathan Beehl wrote: "It appeared in the racing firmament like a Supernova only to disappear almost instantly." Its 8 cylinder engine had a swept volume of only 3 litres, but the two KKK turbos helped it pump out 840 hp in practice and 750 when racing. Its lines were a free interpretation of the "silhouette" formula. The debut of the definitive version of this car was at the 1981 Daytona 24 Hours. It behaved very well during practice and registered the seventh best time, showing a good number of Porsche 935s a clean pair of heels. During the race it was in second place for a spell and

Engine: rear-mounted 90 degree V8, two KKK turbos. Displacement: 2,926.9 cc. Bore and stroke: 82 x 71 mm. Compression ratio: 9:1. Power: 950 hp at 9,000 rpm. Fuel system: two turbos, Bosch injection. Suspension: fully independent. Weight: 950 kg.

managed to record the fastest lap before electrical problems put it out of contention. At the Mugello 6 Hours it made the sixth best time in practice, but it failed to finish yet again due to electrical troubles. Facetti-Finotto's 308 reached 320 km/h in practice for the Monza 1000 Kilometres thus recording a best time which left the nearest rival a good second and a half behind. Unfortunately, during the Sunday morning warm up, the 308 broke down beyond all hope of repair and could not join the line up on the starting grid. Nor did it enjoy better luck at Silverstone, where it again failed to finish. At the Nürburgring, another crash during practice ruled it out of the race. The 308 Carma

made another appearance at Pergusa: pole position followed by withdrawal after three laps when it caught fire. Its last race was the Kyalami 9 Hours. It made the second best time in practice, but even though it again demonstrated just how quick it could be, it nevertheless failed to finish. Facetti and Finotto concluded that the 308 twin turbo was simply too complex for the resources of a private team and they decided to put it down to experience. A real pity.

512 BB LM

Engine: rear-mounted, 180 degree V12. Displacement: 4,942 cc. Bore and stroke: 82 x 78 mm. Compression ratio: 10.3:1. Power: 470 hp at 7,200 rpm. Gearbox: five speeds plus reverse. Brakes: hydraulic discs with twin master cylinders. Frame: tubular space-frame in CrMo steel with elliptical section side members. Suspension: independent, with upper and lower trailing arms, double action telescopic type hydraulic shock absorbers and transverse stabilizer bar. Tracks: front, 1,563 mm; rear, 1,713 mm. Wheelbase: 2,500 mm. Weight: 1,105 kg (curb weight).

The characteristics of this car are summed up neatly in the cypher itself: 5 stands for litres of swept volume; 12 for the number of cylinders; one B is for boxer, the other for berlinetta. The engine is a descendant of the 4.4 litre unit which powered the 365 GT4: the increased displacement was obtained thanks to a new crank shaft which allowed the stroke to be increased from 71 to 78 mm. Born in 1976 as a high performance GT, the Ferrari 512 BB was used for a long time by the private teams involved in endurance racing. Luigi Chinetti was in charge of early testing of the racing

version of the model with the 4.4 litre engine. It was not until 1978, when the 5 litre engine arrived, that Ferrari took over direct responsibility for the preparation of the LM version of the 512, which was then handed over to the private teams run by Chinetti, Pozzi and Beurlys, and Bellancauto. The weight was reduced to 1100 kilogrammes. Power stood at around 460 hp. The aerodynamics of the body were studied with particular care and the rear end was fitted with the same aerofoil as was used in the F1 T3. In the 1978 Le Mans 24 Hours, the four 512 BBs emerged as being very fast, far quicker than the old Daytona GTB, but they all failed to finish due to transmission failure. The BB was slightly modified in 1979 when the installation of fuel injection upgraded power output to 480 hp. The aerodynamics were reworked after experimentation in the Pininfarina wind tunnel. Three cars raced at Daytona, with no luck. Things went rather better at Le Mans where a 512 from the Belgian "Beurlys" team took twelfth place. In 1980 Ferrari put in even more work on the 512 BB: the weight was reduced to 850 kg and miniskirts were added to the lower part of the body, while a beefed up engine produced 500 hp. In the 1981 Le Mans 24 Hours, Andruet and Ballot-Lena produced the BB's best result when they drove one to fifth place. The next year, the 512 BB driven by Dieudonné-Baird-Liber came sixth. In 1983 the organizers of the Le Mans event decided to abolish the IMSA class and the BB had to be pensioned off. The car shown on these pages bears chassis number 38739: it was the last to be built in 1981 and it was never raced.

308 IMSA

Engine: mid-mounted, 90 degree V8, longitudinal. Displacement: 2,926.9 cc. Bore and stroke: 81 x 71 mm. Compression ratio: 11.4:1. Power: 363 hp at 8,500 rpm. Valve gear: 4 valves per cylinder. Fuel system: Bosch mechanical injection. Gearbox: longitudinal, five speeds. Brakes: ventilated discs. Frame: tubular space-frame. Suspension: all independent with unequal length A-arms. Tracks: front, 1,480 mm; rear, 1,610 mm. Wheelbase: 2,340 mm stretchable to 2,450 mm. Weight: 840 kg.

A "mobile laboratory," prepared by Giuliano Michelotto with a view to studying the development of a road car for the Group B races. The engine, which had been mounted transversely on the road version, was fitted longitudinally in order to lower the centre of gravity. A few deft touches here and there enabled the engine to produce more than 360 hp.

The car had a tubular space-frame and a suspension system based on double wishbones. The body was in composite materials. A lot of work was done on the dimensions to obtain the maximum negative lift in comparison with the basic model. The 308 IMSA had a shorter, more compact look. Why IMSA? The name was chosen out of affection for the

stability of the American racing regulations. This car took part in one race only, the 1984 Monza International Rally, which was also open to prototypes. Raffaele Pinto was at the wheel of the 308 IMSA. He was leading before a puncture held him up for several minutes. The race was won by Attilio Bettega in the Lancia, but Pinto's Ferrari nevertheless managed to

finish. Three 308 Prototypes were built at Giuliano Michelotto's workshop in Padova. The first model, number 001, was prepared between 1983 and the opening months of 1984. It was bought by a Belgian driver from the Francorchamps team, Jean Blaton, who had won fame racing under the pseudonym "Beurlys." The car made its first appearance in competition on the 1st of June 1984 at a Ferrari Club sponsored event in Luxembourg.

The second model, number 002, is the one shown in these pages. After the Monza Rally the owner of the car asked Giuliano Michelotto to have the car homologated for road use and it now bears the number PD 748291. The third car, number 003, was bought by Dutchman Henk Koel. Construction began in the Autumn of 1986. This 308 prototype appeared at the Zandvoort circuit on the 27th of May 1987 for a private test session.

GTO EVOLUZIONE

Engine: rear-mounted, 90 degree V8, longitudinal. Displacement: 2,885 cc Bore and stroke: 80 x 71 mm. Compression ratio: 7.7:1. Power: 650 hp at 7,800 rpm. Valve gear: dohc per bank of cylinders, 4 valves per cylinder. Fuel system: 2 turbos, electronic injection. Gearbox: five speeds plus reverse. Brakes: ventilated discs without servo. Frame: tubular steel space-frame braced with elements in composite materials (Kevlar and carbon fiber). Suspension: independent, transverse arms, coil springs. Tracks: front, 1,596 mm; rear, 1,669 mm. Wheelbase: 2,445 mm. Weight: 1,114 kg.

Ferrari put Giuliano Michelotto in charge of the development of the 288 GTO. The project began in 1986. The first laps on the track at Fiorano date from January 1987. The aim was to take the maximum advantage of the possibilities offered by the racing regulations. The project had still to reach its conclusion when FISA banned Group B cars from rallying and so Michelotto concentrated on developing the car's track potential instead. The engine was modified until power output climbed to 650 hp at 7800 rpm with overpressure standing at 1.2 bar (2.2 bar absolute). The compression ratio was 7.8:1. The car burned special 120 octane petrol which was the same blend as used for Formula 1 single-seaters. The development of the F 114 CR 2 engine used for the GTO involved fitting separate electrical systems for ignition and injection, two bigger IHI superchargers, new camshafts with a different profile and increased valve lift, new pistons and trimetal main

bearings. Acceleration was extraordinary: 100 km/h from a standing start in less than four seconds. The gearbox was a racing type. 280 kilos were removed to bring the weight down to a trim 940 kilos. The body was a direct descendant of the 308 IMSA Prototype and was the fruit of a massive effort in the wind tunnel.

The car has numerous air intakes on the wings and wide rear vents. The bonnet and boot open up completely for maximum ease of access to the mechanical members. The driver's accommodation is obviously spartan, but the instrumentation is exhaustive.

Access to the electrics is on the right hand side of the dash and it is possible to vary the braking force distribution between the front and rear wheels. The design of the suspension is the same as that of the production GTO, but the settings are different. Brakes and clutch have recently been upgraded. In 1987 the car was slightly modified in preparation for an appointment at the track: ground clearance was drastically reduced to 70 mm. The outline of the nose was altered. Five models were built, none of them has been raced, and two were used to test the engineering of the F40.

F40 LE MANS

Engine: rear-mounted 90 degree V8, longitudinal. Displacement: 2,936 cc. Bore and stroke: 82 x 69.5 mm. Compression ratio: 7.8:1. Power: 750 hp at 8,500 rpm. Valve gear: dohc per bank of cylinders, 4 valves per cylinder. Fuel system: 2 IHI turbos, electronic injection. Gearbox: five speeds plus reverse. Brakes: four ventilated discs, dual circuit. Frame: bonded tubular steel, integrated with composite materials, Kevlar body. Suspension: independent transverse arms, coil springs. Tracks: front, 1,580 mm; rear, 1,570 mm. Wheelbase: 2,450 mm. Weight: 1,050 kg.

Gianni Rogliatti recalled: "In the June of 1986 someone at Ferrari had the idea of creating a car to commemorate Enzo Ferrari's forty years as a car builder: it had to be a race-bred street car representing the very pinnacle of engineering achievement, the technical synthesis of all the experimentation carried out with research prototypes in a wide range of fields like composite materials, engine design and the manufacture of other components like brakes, suspension and aerodynamics. Today the time required to create a new car is three to four years of study and testing." Despite this the Ferrari engineers broke all records.

On the 21st of July 1987, Enzo Ferrari presented the F40. In the planning stage it had been known as the 3000 Le Mans, in memory of the first rear-engined car destined for the public which had been derived from the 250 P which carried Scarfiotti-Bandini to victory in the 1963 Le Mans 24 Hours. The F40 is a spectacular synthesis of the production GTO and the GTO Evoluzione. The car was built over the 2,450 mm long tubular space-frame which had already been used for the 288 GTO. This was specially reinforced for the occasion with carbon fiber and Kevlar panels. The engine is a 2,936 cc V8 turbo with four valves per cylinder which turns out 478 hp at 7000 rpm with the pressure at 1.1 bar. The body is made entirely of composite materials and the drag factor is barely 0.34 with lift standing at more or less zero. This result is exceptional because it was obtained without recourse to aerodynamic appendages. Work began on the racing version in the January of 1988 with a possible Le Mans entry in mind. 650 hp was squeezed out of the engine with turbo pressure at 1.4 bar. The brakes were modified and discs with larger calipers were fitted. Substantial modifications to the body were not necessary: the F40 had racing in its blood right from the start. Ground clearance was reduced to 60 mm. The dashboard was removed and the Formula 1 type digital instrumentation was installed.